Swan Club

Book of Heavenly Guided Poems

By

Cherubim

Sabrina Naqvi

The story, all names, characters, and incidents portrayed in this production are fictitious. No identification with actual persons (living or deceased), places, buildings, and products is intended or should be inferred.

E-Book ISBN: 978-1-916996-50-2

Paperback ISBN: 978-1-916996-51-9

Table of Contents

Acknowledgements

Amy

In the embrace of Amy's gentle grace,

A heartfelt gratitude finds its place,

Her loving nature, a melody, tender and gentle,

Kindling a flame, dispelling all fear.

Within our unique connection, her presence, patience bloomed anew,

Igniting passions, painting skies in blue.

Acceptance, a gift for my quirky soul,

Unveiling beauty, making me whole.

Her support, a haven, rare and true,

A shelter where dreams find their debut.

For Amy, the guardian of my deep heart's plea, I sing this grateful melody!

Divine archangels, heavenly and bright,

Sent from heaven, in radiant light,

To love, to guide, to heal my soul,

Their presence, a blessing, making me whole.

Charming beings with wings unfurled,

In their grace, my spirit twirled,

Supporting me through night and day,

Their divine love, an eternal sway.

Oh, archangels, so pure, so divine,
Your presence, a gift beyond the line,
Guiding steps, with arid heavenly presence,
In your embrace, I find my heart.
Heaven sent, with grace so fine,
Divine archangels, forever mine.

Oh Heavenly Father, in your steadfast might,
Your love, unconditional, a guiding light,
For your children's salvation, pure and true,
Your unwavering power carries us through.
Infinite love, a divine embrace,
Guiding souls, in boundless grace,
Your strength, a fortress, never to sway,
Ensuring our path, come what may.
Oh, Father above, your love divine,
Rescues our spirits, in every line,
Salvation's promise, in your embrace,
Eternal love, our saving grace.

Jesus' inspired might,
Eternal life, in radiant light.
Through the strength of a lion, love and grace, he shows the way,

He is our hope each day.

His power keeps all dark at bay, eternal, beyond all strife,

Gifts us with everlasting life.

In His name, our spirits thrive,

Through Jesus' might, we revive.

To friends and family, pillars strong and true,

In the tapestry of life, your love shines through.

Through joyous peaks and valleys low,

Your unwavering support, like a steady flow.

In laughter shared and tears embraced,

Your presence, a haven, never to be replaced.

Through sunny days and stormy weather,

Your love strong and true.

In moments bright and those more gray,

Your grace and strength light my way.

For every trial, every uphill climb,

Your love, a constant, transcends all time.

So here's a heartfelt and humble token,

To the bonds unbroken, the love unspoken.

To friends and family, my heart's sweet dove,

Thank you for your enduring, limitless love.

About the Author

Through life's tumult, I found myself bound,
Lost in inner pain, a silence profound.
Standard coping failed to break the chain,
So I sought God's solace to ease the strain.
In the depths of my being, a soul's desperate plea,
Inner child yearning for release, longing to be free.

On the quest for a presence, divine and near,
To witness my heartache, to wipe every tear.
With each step towards the sacred unknown,
Hope blossomed, a sanctuary I'd never known.
In the search for God, my spirit took flight,
A journey of healing, embracing the light.

Through the turmoil, a transformation began,
A dialogue with the divine, a sacred plan.
In the echoes of prayers, my pain found reprieve,
God listened, my heart and soul, in His grace, believed.

My visits to heaven through visions.

119 Heavens they are vast and wide, full of greenery and Meadows filled with sovereign light. With fields of crops and wild stock a place where Heavenly Father goes and tends to Heaven's crops.

The Power of My Heavenly Fathers Unconditional Love

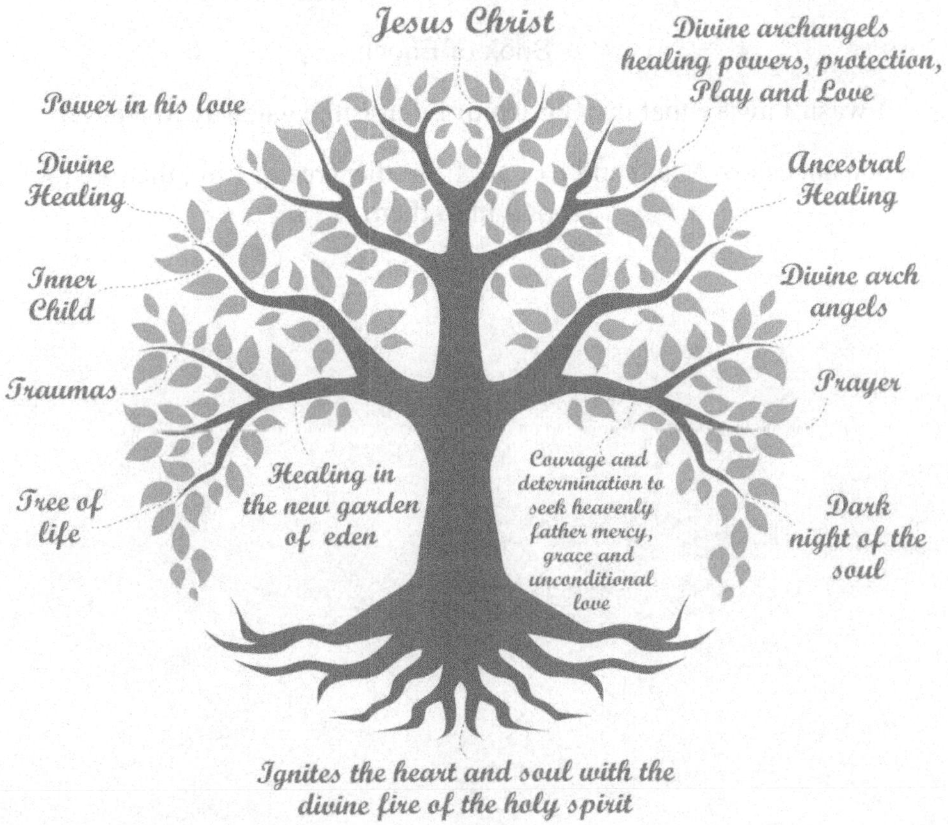

Heavenly Oak Tree

Swan Club

Heavenly Prophetic Book of Poems

"With the greatest power of our Heavenly Father, overcame the evil and sent him a powerful lightning out of heaven into hell."

~ Book of Enoch

I wasn't aware that the Garden of Eden is the gateway to Heaven

When ancestors Adam and Eve ate from the Tree of Life, the tree took root within them.

Dark Night of the Soul

Beneath the veil of shadows,

Ignition sparked within my heart.

In the soul's abyss, a tale did swell,

A journey through the dark night of the soul.

Through Jesus' whispers, a plea so true,

Yearning for grace, unconditional love and forgiveness, bestow anew.

Unconditional, the Father's trace,

Sanctuary found in sacred grace.

Tear Drop Offering and Prayer

In every tear, a whispered prayer,
Each drop an offering, rising through the air.
For in the Lord's embrace, blessings pour,
He reads the tales of every tear's encore.

A story unfolds with each glistening bead,
A sacred dialogue, in prayer we plead.
Heavenly Father, capturing the essence of each heartfelt tear.
In the quiet solace where sorrows meet grace,
A testament of faith in every tear's trace.
May His blessings cascade, a divine embrace,
As the tale of each drop finds its heavenly place.

Inner Child Poem

Oh, dear child within, come close to me,
Into the heaven of your heart, let's roam free.
With tender words and a loving embrace,
I'll wipe those tears from your innocent face.
In laughter and games, we'll find our way,
Building castles in the sand, where we'll play.
The world may be heavy, but you're not alone,
Together, we'll heal, and we'll make a home.
Your dreams, so vibrant, your spirit, so pure,
I promise to protect you, of this, I am sure.
Let's dance in the rain, under the endless sky,
And watch as your fears and sorrows pass by.
With a heart full of love, I'll guide you through,
The pain of the past, we'll bid it adieu.
With patience and care, we'll learn and grow,
Your inner child, forever aglow.

So, my dear, sweet one, please don't fear,
I'm here to love you, hold you near.
In love's disguise, the soul discerns,
As bonds decay, our essence burns.

Yet belief persists, love's sacred embrace,
Our right divine, a cherished grace.
The past, a sculptor, moulds our frame,
Guiding us toward the self we aim.
To confront fears, to openly share,
With love's pure light, our hearts repair.

Face yourself and ancestral roots that need to go,
Within our being, ancestral woes,
Childhood scars, society's throes,
Illusions, delusions, a maze confined,
Traumas seeding fear in hearts enshrined.
Yet amidst this chaos, the soul's serene,
A sovereign light, love's steadfast sheen.
Caged within illusions, fear's illusion cast,
But love, untamed, prevails steadfast.
Amidst the inner storm, its glow persists,

A beacon strong, at the depths of our hearts where the truth exists.
In hearts, souls, and minds confined,
Love's sovereignty eternally aligned.

Heavenly Father, Jesus dear,
Protect, guide, let no fear near.
Lead me towards the sovereign light,
In our hearts and souls, where Heavenly Father's victory takes flight.
Grace of Heavenly Father's embrace,
Unconditional love, a sacred space.
Guide me, lead me, oh, divine above,
Towards the victory unconditional.

In shadows deep, a graveyard sprawls,
Humanity veiled, akin to thralls.
Like zombies roam, their eyes obscured,
Illusions bind, ancient fears endured, awaiting judgement.
Since ancient times, at fear's grasp,
By Adam's and Eve's fall from grace,

In Eden's timeless place.

Society weaves a veiling shroud of deception and lies,

That keeps you all bound, as the unseen battle rages on.

Conditioned minds, resistance gone.

Yet deeper still, within the soul,

A healing balm, can make us whole.

Heavenly whispers, messages divine,

Await those who seek, in hearts entwined.

Awaken, break the chains that bind,

The conditioning of the mind.

For in this struggle, truth shall gleam,

And from the haze, our spirits redeem.

Love, wisdom and heavenly power, unconditional love of our Heavenly Father through Jesus, the son of our Heavenly Father,

Jesus had an intimate and unconditional loving relationship with our Heavenly Father in the Holy Spirit.

Confront the darkness, inner fears and ancestral traumas

Purge the heart and soul to allow the light and unconditional love to flow.

Open your hearts to Heavenly Father and with his divine Angels and unconditional love shall heal our hearts, minds, body and soul.

In the embrace of Heavenly Father's grace,

Unfolds a dance with Angels' gentle and strong protective embrace.

Unconditional love, a soothing balm,

Heals the hearts, minds, body, and soul's calm

Open wide heaven's gates, let love flow,

His tender touch, a healing glow.

With divine Angels standing strong close by

Hearts unite, casting out fears to his delight.

"I tell you, not seven times, but seventy-seven times." – Jesus

"If you forgive others their transgressions, your Heavenly Father will forgive you." – Jesus

In the heart's deep embrace, forgiveness blooms,

Beyond earthly rooms.

Unconditional love, a heavenly guide,

Unveils the path where healing resides.

To forgive the self and others dear,

Unravelling knots, shedding every fear.

Journey within, with intentions pure,

With love as the compass.

Grant yourself space, time to reflect,
Ponder what's needed and may wisdom gleam.
Forgiveness unfolds, a divine stream.
Inner and ancestral wounds set free,
From the heaven's unconditional love transformative power, heals all.

In a world of flawless quests,
A good heart outshines the rest.
Perfection's charm may fade away,
But heavenly love lasts, come what may!
Your good hearts are the real work of art,
We are made from our Heavenly Father's unconditional love.

Jesus appeared to me in the waiting room of Heaven
He was so bright he stood in a white robe and in heavenly pure white light.
I rushed toward him and we embraced each other.
We held hands and raised our hands up and declared joyfully his victory.

In each beating heart,
There lays a seed of unconditional love.
In every heart reveals its truth,
Within every soul its story
That only Heavenly Father can bear witness to,
Through his only begotten son, Jesus.
A beacon of bright light.

"Sticks and stones can break your bones
But words will never hurt me,"
A delusional saying that many generations
Misleadingly taught their children to believe.
I'd prefer sticks and stones over the venomous words
Wielded by others
That have the power to pierce and attach to the heart and soul,
Leaving lasting scars upon precious hearts.

In the haven of the Garden of Eden,
Where divine energies intertwined,
A tale of profound healing unfolded.
My essence traversed realms,
Guided by the ethereal touch
Of angelic healer Archangel Raphael.

With resolute courage
And the fiery embrace of the Holy Spirit,
I unearthed the Tree of Life,
Entrusted by ancestors to safeguard it,
Warding off malevolent forces.
Through fervent communion
With our Heavenly Father,
I beseeched for unconditional love,
Healing, protection, and the liberation
From inner afflictions.

Archangels Raphael and Michael,
Graced my presence,
Their heavenly presence accompanying me
As I embark upon the journey within Heaven's garden.
Here, amidst a crystal spring lake,
Archangel Raphael, Heaven's great healer,
Performed profound healing,
While heavenly angels rejoice.
However, beyond the garden's sanctum,
I confronted a foreboding gateway,
Shrouded in darkness,
Presiding over a cemetery of veiled souls—
Humanity obscured by ancient blindness.
There stood to the right of Heaven's gates,
Something dark, obscure,
Wielding dark incantations,
Ensnaring lost souls.

In a clash transcending earthly realms,
With the great might of our Heavenly Father,
I battled victoriously
The malevolence standing right of Heaven's gates.
This was not a battle against mortal flesh.

A thin slip in Heaven's gates appeared to me.
Ascending the staircase of the 7th Heaven,
Unwavering love, victory and rejoicing among the angels.
Standing to the right of the heavenly throne Jesus,
Our clasped hands raised into the heavens.

The lineage of Judea, stood Jesus, King David and King Solomon and me.
Jesus let me climb the stairs
Where a beautiful white and divine lion stood waiting.
I ran towards the lion exclaiming, Heavenly Father!
I was so excited,
I wrapped my arms around the lion,
The lion looked over at me and smiled.

Heavenly Father showed me,
In my dreams,
A very divine baby and the word Chraub.
Once again, I didn't know anything
About this word or what this divine angelic baby was.

I researched to discover that Chraub
Is a divine presence in God's heavenly kingdom.
Chraubs are throne-bearers
And were appointed to guard the Tree of Life.
Their role is
To play intermediaries
Between heaven and Earth.

A dove presented to me,
Which when I read the significance of the dove.
It means the divine unconditional love
Of our Heavenly Father
Is present and communicates with me.

I heard in my heart Jophiel.
Once again, I had never heard of this name.
I looked her up in the Bible.
Archangel Jophiel is beautiful
And full of wisdom
Who at the beginning of creation
Spoke 70 languages into souls.
She has much wisdom, grace and love.
She is also the Prince to Cherubim,
A heavenly divine Angel.

Sky Blue Banner - Lion Tribe of Judea

In the realm of the 7th heaven, we rejoice,
Banners of sky blue, lion tribe of Judea.
Together, we victoriously battled against forces of the night,
Embracing the power of unconditional love, dispelling all blight.

7th Heaven

Heavenly Father's Powerful Army Resides.
The army of angelic Angels ascend a heavenly staircase,
Where the angels bow.
We sing, dance in praise,
Love, and rejoice
In the unconditional love
Of our Heavenly Father's delight
That shines upon his children.
In Heaven's sovereign light,
Heavenly Father's voice
Paints the night.
In every heart
The seed of truth resides.

In foresight's gentle breath, ancient wisdom speaks,
An angelic whisper, where divinity peaks.
I was led to following biblical passage Jericho –

From the Torah,

"Jericho's mystery, a script unfurled, Guided by 7th heaven, in an enchanted paradise. Seven days they marched, at divine behest, Conquering walls with faith's unwavering crest. In the seventh realm, angelic wings take flight, Angelic angels bowing to love's might, in heaven's light."

At Heaven's throne, Rejoice with Father, in realms divine. Archangel Jophiel and hosts unite,

In conquering darkness with love's pure light.

The Chicken or the Egg

In a realm where mysteries twirl and dance, lays ancient riddle that leaves us in a playful trance. The chicken or the egg, which came first, we say, But in the whispers of the heavenly Divine, the answer may sway.

With a twinkle in His heavenly eyes and a chuckle so divine, The Heavenly Father shares His love, oh so divine. For in His wisdom, He knew from the start, The chicken would emerged from the egg, a masterpiece of art.

Yet nestled within, in a delicate shell, Lay the promise of life, a story to tell. So, in this waltz of creation, we find, The chicken and the egg, forever intertwined.

Let's raise a cheer, let laughter ring, For in this puzzle, joy does spring. Oh, what delight, in this revelation of Divine creation!

7th Heaven: Heavenly Fathers Army Resides

An army of angelic beings ascends a heavenly staircase, reaching the 7th Heaven where we humbly bow to our Heavenly Father,

Within this realm resides a powerful heavenly army.

We sing, dance in praise, love, and rejoice in the unconditional love of our Heavenly Father.

As we find joy in Him, His delight shines upon us in return.

Spark in the Night

In the quiet of my room, with a prayer sincere,
I spoke to Heavenly Father, the lion, so near… through Jesus.
A cover pulled back, a heavenly light so bright,
A spark from heaven, filling the tranquil night.
Startled, I stood, in awe of the divine heaven,
Laughter echoed, a joyous, heavenly sign.
With archangels around, a chorus
In that sacred moment.

Heavenly Father and I Play

I would slide into my bed with my hips on the mattress and a beautiful heavenly sparkling light would appear.

Startled! I stood in awe of divine heavenly presence.

Along with the Archangels around,

A chorus in that sacred moment echoed.

Eve and the Tree

When Eve went missing, Heavenly Father knew she was at the Tree of Life and sent a cherubim in the full armour of God, sword in hand to the Tree of Life. Cherubim reached the tree to witness Eve lifted two inches off the ground, she was pale. Cherubim charged towards the tree, sword raised and sliced the evil in two pieces and Heavenly Father sent a Heavenly and powerful bolt of lightning into the tree, hurtling the tree onto the Earth and within it deep.

In life's grand script, doctors diagnose with care,

Prescribing remedies for "conditions" we bear.

Societal and family conditions, a delicate thread,

Where emotions are spread.

Experiences shape us and take root,

In the garden of relationships, emotions grow.

Conditions diverse, like a tapestry,

Yet through them, we glow.

Abu Dhabi

Three religious sights in Abu Dhabi,

Unique religious buildings that reflect

A different spiritual path,

Embracing diversity in the tapestry of faith

And yet not on holy land.

While some may find sanctity in other places, Jerusalem, Israel, stands as a timeless and ancient beacon from the heavens, rich in history, the lineage of Judea and revered as the true pillar of worship across traditions.

Heavenly Rapture

In rapture's grasp, a dance unfolds,
Exhilaration mixed with tales untold.
Scary whispers of the great unknown,
In Heavenly embrace, seeds are sown.
Lord Jesus, guide us through the haze,

Heavenly Father, in Your ancient wisdom, we gaze.
Rapture's sensation, both wild and free.
A heavenly waltz, a soul's and heart's jubilee.

Heavenly Divine Revelation

Enemies of humanity and Israel coming from the western leaders.
America, Russia and North Korea. Will send two missiles from either land
toward the Middle East. Two leaders in the Middle East will take refuge in
the western land after betraying their own holy land of the Lord and
Savior, Jesus.

In Skies

Heavenly Father, through a vision of a dove. Jesus,
Lord of Lords, King of Kings,
Saviour in the eastern sky, where the sunlight sings.
A chariot divine, with two stallions
Pure heavenly white glistening in gold.
Appearing like the Sun in royal array.

Visions of A Dove, Lion and Eagle

Those that make it through the ajar Heaven's gates
Shall climb the stairs of heavenly truth of my Heavenly Father.

Heavenly Father awaits at the top, on His throne
With an eagle on either side and Jesus To the right of His throne,
Lord of Lords and King of Kings,
Jesus' throne has two heavenly sovereign lions.

Heaven glistens sovereign light ~ Cherubim

Mother of Israel

In lands of ancient tales and sands,
Where twelve tribes once joined hands,
Children of Israel, proud and bold,
Yet now a tale of greed is told.

Across the holy land they roam,
But acts of greed now fill their home.
Genocide upon Palestine's shore,
A sight that makes my heart grow sore.
For in the ways of Heavenly light,
Such actions dim the sacred sight.
Jesus taught of love and grace,
Not strife and greed in any place.

Let us return to virtues old,
To kindness, peace, and stories told,
Of compassion, where hearts unite,
And darkness fades in love's pure heavenly light.

Heavenly Father's Word

"With the greatest of our Heavenly Father's power overcame the evil by sending a bolt of lightning from the heavens casting out the evil from heaven."

~ Book of Enoch

I wasn't aware that the Garden of Eden is the gateway to Heaven's gates until I witnessed it for myself.

~ Cherubim

When Adam and Eve ate from the tree of 'life', the tree of life took root within them.

~ The Bible

I was wearing the full armour of our Heavenly Father's protection outside the gates of heaven.

~ The Bible

Facing the graveyard of enshrined souls, with Archangel Michael's divine mighty wings.

~ Cherubim

We are children of our Heavenly Father, created from His and only His, unconditional love.

~ The Bible

Fathers Words & Visions

In the heavenly kingdom of Judea, where skies embrace, Heavenly Father reigns with eternal grace. His locks, a cascade of purest white, Glisten like stars in the softest light. With beautiful sovereign eyes of gold that shimmer and shine, He watches over all, a love divine.

A smile so vast, it lights up the skies, Guiding children's hearts and souls on their heavenly rise. On his throne above, in realms so high, He sits between eagles and angelic angels, majestic and wise. Jesus reigns as King in the Earth and heavens and in the seventh heaven, awaits as love and compassion through eternity gates. In this heavenly kingdom, eternal love and victory thrive, where love so pure and heavenly crystal clear spring water flows, forever my children.

Oh, Heavenly Father, in eternal splendour and light, Your presence fills our hearts, minds and souls with pure delight.

I was guided to the Swans. I was then guided to the Bible and significance of swans.

"The Seven Swans, a Swimming verse, represents the seven gifts of the Holy Spirit (Romans 12:6-8). Prophesy, service, teaching, encouraging, giving, leadership & mercy. Children were taught that when you walk with God, the gifts of the Spirit move in your life as easily as a swan on the water."

Whisper and Foresight

Archangel Michael proclaimed with vigour, "HOLD THE VINE!"

The Army of Heavenly Archangels of Judea walks with Yeshua.

We conquer the darkness in the heavenly realm.

I walk with an army of Heavenly Angelic Angels, divine noble warriors.

Where the Heavenly Father communicated to me through my heart and visions. Initially, the message was vague, but it gradually became clearer, emphasising the phrase "Hold the Vine".

Heavenly Father's guidance led me to explore sections of the Torah, particularly Sirach 24:17, which links the vine to the wisdom of my Heavenly Father. The Torah depicts Divine wisdom akin to a vine, symbolising the blossoming of delights and the flourishing of abundant fruit, the growth and richness associated with great heavenly wisdom.

Divine Gifts

I received divine gifts from the Heavenly Father: foresight, vision, the ability to communicate with the seed that lays dormant.

Messages from Heavenly Father

Sovereign Sun, Moon and Stars

I sleep in The Heavens among the stars, comforted by pure heavenly Angel wings.

Message from a vision of Heavenly Father on His throne, in the heavens.

"Jesus stands in the heavens outside the gates of the 7th heavens, two sovereign Lions on either side of him, representing His throne. King of Kings, Lord of Lords Jesus, saviour of hearts and souls."

"With Heavenly power, I shall cleanse the sins of the hearts and souls, spanning ancient times, and nourish the mind with my spoken words."

"In the beginning, there was no religion, merely Heavenly Father's unconditional love, whom takes delight."

"I proclaimed to the heavens with might and valour 'Let's make a blazing path for the return of Jesus'."

"What restores in the Kingdom, shall bring judgement onto the Earth."

The Heavenly Father guided me to talk about Archangel Raphael and his incredible healing powers bestowed upon the Earth and the heavenly realm. This heavenly archangel is referenced in various religious texts like the Quran, Bible, and Enoch, showcasing his significance. Additionally, Archangel Raphael is said to sound the trumpet.

Heavenly gates are accessed at four entry points, and guarded by four archangels. Heavenly Father divinely communicated to me, marking my entry into the arch of heaven gates on the Eastern Sky, where Archangel Raphael guards. Heavenly Father had me place heavenly divine angels, soldiers wearing the full armour of our Heavenly Father. Solid sovereign gold, titanium and big in stature at the homes of loved ones, including Amy's home. Heavenly Father, "These are the homes of my children".

"THE VINE IS STRONG"

Heavenly Father's boundless, unconditional love, victoriously shining divine light. Ushers in like a wave from the ocean over the resting place of His children, souls standing outside the gates of the 7th Heaven. He showers His might upon the forces of evil opposing Him and His heavenly offspring. Embrace the summons and wield authority through the path illuminated by Jesus, from the East.

In a charmingly powerful revelation, My Heavenly Father revealed in his loving grace, I stood side by side, embracing powerful authority, hand in hand, cloaked in the full armour of our Heavenly Father's unconditional love, protection and sovereignty. With all mighty powerful presence of my Heavenly Father's passion for his children. Archangel Michael resolute with us, at Heaven's gates on the Eastside of the gates of the 7th Heaven. Further East of where we are standing in our Authority bestowed upon us by our Heavenly Father, towards the mountains beyond the horizon, as the Sun rises the Messiah on a large Chariot, and two heavenly sovereign stallions appear.

"In the tumultuous north, two tribes wage war with man-made weapons of destruction, wielding missiles and drones as instruments of chaos.

Their objective, Israel, the Holy Land, and create an inhospitable realm for the awaited messiah's return.

Behind the scenes, a collaboration unfolds within the Israeli political arena. Two figures, harbouring ulterior motives, conspire to orchestrate an attack on their own homeland, seeking refuge and redemption in a distant land".

The Word Authentic

My Bed

Upon my bed one evening's grace,

Rosary overhead, my room a sacred space.

Above my headboard, a message divine,

Guided by the heavens, in whispers entwine.

Jumped up on hands and knees,

Guided by a compass, the cross to appease.

North, the top, where the journey did start,

East, then South, and the West played its part.

East of the crucifix, Heavenly Father's guide,

A sacred moment,

Falling on my bed, joyous laughter from the heavens,

A connection so divine

Heart of Mary Appeared in Front of Me

Heavenly Father presented in front of me a vision. Jesus and I embraced, as we ascended into the heavens. Jesus and I, surrounded by the heavenly divine sovereignty of our Heavenly Father's army, the Heart of Mary was presented as we ascended into the heavens.

Your Messiah, resplendent on a white stallion adorned in pure heavenly white, sparkling gold, gallantly returns. Leading our mighty army of archangels, we illuminate the sky in divine purity, charging forth with absolute authority toward the pit of deception and illusion that ensnares my beloved children. Together, we usher in the dawn of judgement, breaking the bonds of evil. In this Heavenly divine revelation from my Heavenly Father, the word 'Authentic' echoes through ancient times.

"The convergence of nations — America, Russia, Israel, and more, signals the prophesied end times, marking the demise of the anti-Christ."
~Authentic Word

In the heavens, trumpets sound, the triumphant return, the true Messiah, King of Kings, Lord of Lords, has secured victory, heralding joyous celebration within the Kingdom of Heaven, Crowned and taken up his rightful place on the Heavenly Throne. Heavenly Father - Heed my words, children; embrace repentance, for judgement looms, and only through the lineage of Judea, the Tribe of the Lion, under the banner of sky blue, can one enter the gates of Heaven, through my beloved Son, Messiah, King of King, Lord or Lords.

Swan Club Love

Two peas in a pod, side by side,
In life's garden, they do reside.
Together they grow, in harmony,
Like a perfect pair.
Their connection strong, through thick and thin,
In laughter and tears, they always win.
Two peas in a pod, forever entwined,
A bond so deep, one of a kind.

Worth It to Get to You

Through trials and tribulations, we tread,
Enduring sorrows, where shadows spread.
For in the darkest depths we find our way,
To reach the light of a brand new day.
Each challenge faced, a lesson learned,
With every scar, our hearts have yearned.
To find the strength within our soul,
To mend what's broken, to make us whole.
Sorrow may linger, troubles may persist,
But in our hearts, we must insist,
That through the storm, we'll always strive,
To keep our dreams and hopes alive.
For in the end, when the battles are through,
We'll emerge stronger, tried and true.
Tribulations may test us, but we'll endure,
To find a love and life that's pure.

Made in United States
North Haven, CT
10 November 2026

22091349R00029